Cultivate Your Audience

Attract Ideal Buyers, Connect With Your People, and Create Consistent Income in 90 Days or Less

Lise Cartwright

Disclaimer

This book is for informational purposes only.

Some of the information found within the contents of this book is about third party products and services. These Third Party Materials consist of products and opinions expressed by their owners. As such, the author does not assume responsibility or liability for any Third Party material or opinions expressed.

The use of recommended Third Party Material does not guarantee any success and/or earnings related to you or your indie author business. Publication of such Third Party Material is simply a recommendation and an expression of the authors' own opinion of that particular material.

Links to Third Party Resources may be affiliate links, meaning the authors may receive compensation if a service is ultimately purchased from such a link.

Contents

YOUR FREE GIFT!

To get the best experience with this book, get instant access to the Resources Hub, where you'll find additional templates, trainings, and resources to help you *cultivate your audience* with ease.

You can grab a copy here:
www.hustleandgroove.com/cyaresources

Introduction

I couldn't open the email.

What would they have to say? Maybe they're going to tell me how crap my book was... I'm not prepared to hear that from a complete stranger.

Or maybe they are going to say how dumb I am and that I should NOT quit my day job!

What am I meant to do? Everyone is telling me that I need to build my email list now!

But why? That's what I don't understand...

Meet Shy Sue. These are the thoughts she has around using email in her business. She's a non-fiction author and course creator who loves writing and sharing her message with the world as long as she doesn't have to respond to emails or create landing pages, or any of the other tech stuff.

Shy Sue just wants to create.

Shy Sue has a dream. A big dream. She wants to be the next Oprah Winfrey or Rachael Hollis, impacting millions of women around the world through her words and teachings.

Only Shy Sue has no idea how to get her ideal audience to join her email list, nor has she been able to figure out how to create one of those landing page thingees...

Nope. Instead, Shy Sue continues to focus on her writing and hopes that one day, a magical publisher will knock on her door and all her dreams will be answered...

Oops...

Shy Sue just woke up from her daydream and realizes that if she is going to spread her message and reach millions of women, it is going to take some action on her part.

But where to start?

Hi! I'm Lise Cartwright, self-confessed shoe fanatic, non-fiction author of 40+ best selling books and creator of the List Building Collective, a membership site and support hub for creative entrepreneurs who want to grow their email list and make money through emails — in as little as a few hours per month.

Over the past 10 years, I've been honing my offers and connecting with my email subscribers.

Every day, I receive emails from new email subscribers telling me how much they love my energy, my positive vibes, and the knowledge I so willingly share through my books, courses, YouTube channel and membership site.

But it wasn't always this way.

Oh no. If you met 2014 Lise, you'd be forgiven for thinking you were chatting to the wrong person!

2014 Lise was shy, scared, timid, and not particularly tech-savvy.

'Present-Day Lise', on the other hand, is a chatty, energetic person (when she's aligned), always the first to put up her hand to help, and dangerous enough with her good friend Google at her side.

Your version of this might look different, but if you want to connect with your audience, build an email list of loyal followers, and have raving fans ready to share your message and buy your things, then you have to step out of your comfort zone and approach things a little differently...

...based on what that looks like for you.

WHO THIS BOOK IS FOR

This book is for you. The creative business owner looking to make a real impact in the world.

It's for the non-fiction author who longs to hear her message being spread around the world.

It's for the course creator looking to transition into delivering coaching or digital products.

It's for you if you have a desire to change the world, leave a positive mark, and feel good about the work you do every. single. day.

WHO THIS BOOK IS *NOT* FOR

My friend, if you're reading this and you already have a successful online business with an audience who loves EVERYTHING you do, then you're not likely to learn much from these pages.

Sure, you'll get a kick out of imagining what my NZ accent really sounds like as you flick through these pages, but that's about it.

So rather than waste your time, go ahead and stop reading now and come and join us inside the List Building Collective instead: .

If you've already purchased, then ask your online retailer for a refund or gift it to a friend who needs it.

You deserve to use your time on what actually benefits you.

IT'S YOUR TIME TO SHINE

Still with me?

Awesome. I'm excited to dive into the pages of this book and share with you the simple, easy-to-learn way of connecting with your audience, nurturing them like a newborn baby, and wowing them so they stick around forever like they are your BFF.

I'm excited to get started. Honestly. I'm grinning from ear-to-ear as I write this, bouncing in my chair, waiting for you to dive into the first section.

So don't leave me hanging!

Come join me in *Part 1: Amplify Your Reach*, where we're going to start attracting your awesome audience.

It's your time to shine, impact, and change the lives of your ideal leads, customers, and clients!

PART 1: AMPLIFY YOUR REACH

"The early bird catches the worm..." ~ **English Proverb**

In this first part of the book, we're going to look at how to capture the attention of your ideal audience.

What makes them stop in their tracks? What makes YOU stop in your tracks...?

Now, before we go any further, I want to assure you that what I'm about to share with you is neither icky nor 'black hat' in any way.

You should treat your audience with the respect and kindness they deserve.

So under no circumstances are you trying to snare them with some gimmicky gifts that don't serve a purpose for your audience.

We are not one of those infomercials!

So what am I talking about here? I'm talking about figuring out what your audience wants right now and fulfilling that need with a digital free gift or low-cost offer (something that's under $20).

You might hear this referred to as a lead magnet, freebie, opt-in incentive etc.

They are all the same thing. An incentive, an offer that your audience wants and they are willing to hand over their email address (or dollars) to access.

That's what we are going to dive into first.

Finding the best audience attracting offers for your potential email subscribers.

Because without this piece, no amount of "join my newsletter" is going to be enough to entice them to hand over their email address...

Step 1:

Choose an Offer

Attracting the right people (leads) into your online business is about creating the right lead magnet (also called a freebie, free gift, incentive) or low-cost offer that provides value to your potential client or customer.

Your ideal person needs to see your lead magnet and go, "*Wow! I have to have this. I can't believe it's free!*" or "*I can't believe this is only $9. I'm buying it!*"

How do you do that?

You figure out what your strategic lead magnet(s) are and create them!

What the heck are strategic lead magnets Lise?

A strategic freebie are offers that are going to bring in the right people who want what you have to offer.

Your job is simply to make it easy for your audience to self identify, for them to raise their hand and say, "*Yes. This is me, and this is why I know this is going to help me.*"

In exchange for providing their email address, they receive your strategic freebie aka the free gift they have to have.

Now all going well, and in a perfect, happy-creative-business-owner world, every visitor that lands on your website will grab your strategic freebie straight away—happy to hand over their email address...

But unless your strategic freebie is the *best lead magnet for that person in this moment* in time, they can end up leaving your site without joining your email list. And that's ok. You don't want everyone to join, you just want your *ideal* audience to sign up.

Which is why I want to highlight that a few things need to align before someone hands over their email address...

#1: Your strategic freebie needs to be the thing they need right now. It's a case of meeting them where they are based on the problem they believe they have. There's no point someone joining your list and not being able to implement your thing.

#2: Your strategic freebie also needs to be something that your potential person feels they can take action on right away. If it looks like too much effort, they'll skip grabbing it, meaning they'll skip joining your email list. Quick wins are what you want to focus on helping them achieve.

Which leads to the question: *How can you find your best strategic freebie for your ideal audience?*

Let's work that out by coming up with different ideas. And let's focus on ideas that naturally lead people into the next step we want them to take... to buy something from us.

We also want to keep it simple.

THE 5-PART FRAMEWORK TO YOUR BEST LEAD MAGNET

#1: Brainstorm

I know this sounds kinda simple, but I've always enjoyed making things simple, and brainstorming is the best way to unlock ideas you hadn't considered.

So to get started, grab a piece of paper or whatever you do to take notes, and set a timer for 15 minutes. I find the timer on my iPhone works best.

Here's what you're going to do:

Write down as many ideas as you can for lead magnets that you feel would be the best fit for your ideal customer/client.

A few things to consider before doing this:

- You should think about high-value ideas, like courses, video training, etc.

- Swipe files, templates and checklists are also great ideas

- Avoid ebooks on their own, they have lost a lot of their value... unless they are an existing published ebook, then it could work

Free-write for 15 minutes until you have at least 20 ideas.

Remember, you're focusing on ideas that would work for YOUR ideal audience but also the types of clients/customers you WANT to work with.

#2: Check out your competition

The next step is to see what your competition is doing.

Find your 5-10 closest competitor's websites and check out what they're providing in the way of lead magnets.

Sign up for their lead magnets and see what their autoresponder sequence looks like too.

Caveat: DO NOT COPY THEIR LEAD MAGNET! Instead, think about how you could make it better, provide more value, add to it etc.

Filter your competition's lead magnets through your own goals and purpose:

- What is your ultimate goal with your business?

- Are you selling services or products?

- Are they direct competition or just in a similar niche? If direct, be careful about what you produce, again NO COPYING!

By now you should have a large list of lead magnet ideas, adding more from this competitor research.

#3: Survey your list

Have you ever asked your potential customers/clients what they actually want or need?

I have to admit, I was pretty scared about sending out a survey to my Hustle & Groove email list—until I did it that is!

It was such an amazing experience. I had a lot of responses and most people were happy to answer a few simple questions.

This has led to me being able to provide even more value via my Hustle & Groove emails and private Facebook Group.

So if you haven't surveyed your list, this is something you should do now.

Don't have an email list? For those of you who are just getting started with your creative business or email list, you might not think that you can do this step.

But you can.

Instead of sending an email, simply share your survey on the social media platform(s) that your ideal potential customer/client is hanging out on.

No excuses! You have to do this step if you want to identify the best lead magnet to attract the right leads into your business. Otherwise, you're going to end up with a ton of leads who aren't the right fit, making your business no fun.

This is what it looks like when the wrong leads end up on your email list...

Argh! Why is no-one buying my services? They downloaded my lead magnet, but none of my subscribers seem at all interested in buying my things.

I get comments like "I love your stuff, but have no money." and "I'm too busy." or "Next paycheck, I'll buy."

I have over 1,000 subscribers but not one of them wants my thing. I just don't get it.

This is what we are looking to avoid.

There are a number of reasons why you can end up in this situation. One of those is not understanding your audience and what they want.

Another is providing them with a product or service that they aren't ready for yet.

This is why it's important to survey your target audience and understand who they are and what they want and need.

Never assume you know best.

So coming back to the survey... The most important question you need to ask inside your survey is this:

"What's the #1 biggest challenge you're facing right now in/with [insert your topic here]?"

The next few questions can be things like:

"What would you like to see more of on the site?" *(multi-choice, provide them with options!)*

"Where are you in [your topic here]?" *(multi-choice, starting out, intermediate or advanced)*

Make it easy for people to respond and keep it short! No more than five questions is ideal.

Compile the answers and add any new ones to your growing list.

#4: Rank your ideas

Now that you've got a huge list of potential lead magnets, it's time to narrow things down.

Remember, you're looking to provide a lead magnet that is high value. In fact, it should be something that your ideal client/customer would pay for.

In fact, you don't have to a freebie. You can take any of these ideas and turn them into a low-cost offer that also attracts your ideal audience into your world.

It simply depends on what the next step is you want your new email subscriber to take.

For now, let's assume you're going with something free.

Some great examples include:

- A resource library (perfect for you if you already have a lot of resources you use in your business)

- A video course (3-5 videos is the sweet spot)

- Swipe files (email copy works great here)

- Templates (providing them with a template with (video) instructions is a great idea)

- Cheatsheets (great if you've got some special html code that you use to get your site to do something special)

#5: Identify your top three

Identify the top three ideas and then create all three of these lead magnets.

Why limit yourself to one lead magnet right? But know this — if it feels like a lot to create right now, then just do one first. You can do more later.

In fact, if you want to be a smarty pants about it (who doesn't?!) then you should create lead magnets for the different levels that your audience might be at (achieving the 'meet them where they are at' idea):

1. Beginner, just getting started

2. Started, looking to grow

3. Advanced or nearly there

By doing this, you'd have different funnels that would lead into your products or services that meet those needs.

Then you could upsell/downsell based on the actions they took during the email sequences.

Don't worry about this now though. Get the free gift created then worry about the tech stuff later (we'll cover this later in the book too).

But what about the ideas you have left?

With the list of ideas that you've got left, consider which ones fit within your business model and your overall strategy for your creative business.

Make a plan to review these ideas against the ones you create and see if you can create another for your best-selling product or service... that way, you're offering more ways for someone to get into your funnels!

Alternatively, you could also turn those ideas into low-cost offers. Something I'm a huge fan of doing if I have a new program launching soon.

When people pay for your low-cost offers, they are more likely to consider your other offers too.

Keep this in mind as we move into the next chapter.

Step 2:

Capturing Your Leads

Now that you've created your best lead magnets, we need to capture those leads in some way.

But how? I'm glad you asked.

The solution is to create a page where your new leads can go to give you their email address and, in return, they get your amazing free thing aka your best lead magnet(s).

Now, in order to do that you need to give them a place to "opt-in" and give you their email. This is where your website becomes very important because it creates a way for your website visitors to opt-in to the free gift you're offering.

Don't have a website? Not to worry! I've got you covered. We'll talk about your options in a moment.

Now, it might seem super scary to take this step... allowing a way for people to get on your email list. But it's key to your ongoing success.

You want to work with the right clients or customers, right?

You want to automate your income, yes?! Consistent income starts with attracting and capturing the ideal leads. That's what we're focusing on here.

Then we both agree. You need to master this opt-in process.

If you want to help yourself sell more of your amazing thing, you need to be consistently building your email list full of potential customers and clients.

When you build your list and then treat people on your list with care and respect, they will be excited to hear all about your projects.

I want to cover exactly how you can design, write and set up your opt-ins, but first, let's take a deeper look at the types of opt-ins we're going to cover in this book:

- Opt-in Form

- Opt-in Box

- Landing page

Let's get started.

UNDERSTANDING THE OPT-IN PROCESS

I want to cover what each of these opt-in pieces are right now. Later on in the book, you'll actually learn how to create these either through your email service provider (for those who don't have a website yet) or with some of the tools and plugins I recommend.

#1: Landing Pages

Now, let's talk about landing pages. In basic "marketing" terms, a landing page is a single page that has one specific and defined objective. It's also the glue that brings all things into one place for capturing your leads. Your opt-in form lives on this page.

The cool thing about a landing page is that it gives you a few options.

You can set up a landing page on your own website.

Or you can set up a landing page using your email service provider.

You simply point people to the landing page URL to sign up for your email list and download your lead magnet.

It doesn't matter where the landing page is hosted, as long as it provides a place for your ideal lead to leave their email address, you've won.

Anyone who visits your landing page has to make a choice...

They opt-in to grab your free gift or they leave without providing their details.

The goal is to provide a great hook to entice your potential lead; a great graphic that displays the free gift and a way for them to access it.

The decision should be easy for your ideal lead (and work out in your favor).

Here's a few examples of landing pages:

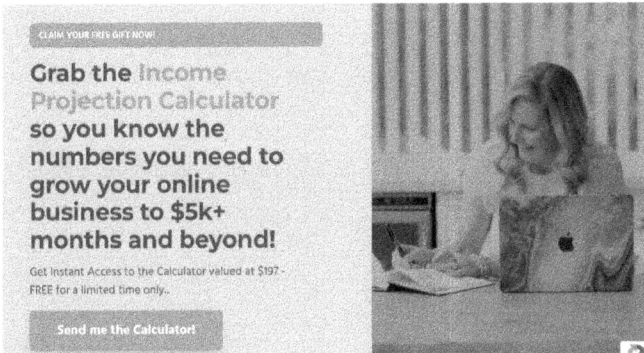

#2: Opt-In Form

Chances are if you've spent any time browsing some of your favorite websites or blogs, you've come across an opt-in form.

An opt-in form, also sometimes called a subscriber box, is a small area on a website that asks for the website visitor to sign up for a mailing list.

These forms can take all sorts of different forms. They can be a popup, they could be something that slides in at the bottom of a page, or they can be a bar across the top of a website.

There are a number of different plugins you can use to create an opt-in form for your website, or you can also create them through your email service provider (both of which I'll cover a bit more in-depth later on).

While you don't need to put extensive time and effort into designing them, you do need to couple it with your best lead magnets.

Here's what an opt-in box pop-up might look like:

Need help with your emails?

You came to the right place

Your email address

Submit   No Thanks

At the end of the day, your goal is to create a landing page to showcase your lead magnet.

CREATE YOUR LANDING PAGE

Let's recap the basics you need to have: a good opt-in form, a landing page, and an opt-in box (if you want to go all-in).

The landing page is a stand-alone page and on it will appear the opt-in form, along with other information to sell yourself and/or your product.

It's very important to remember when it comes to your landing page you don't need to overwhelm visitors with a ton of information.

First, all you want is enough information to get the attention of your ideal lead.

Second, to show the visitor that your lead magnet is amazing and will help them.

Third, to get the visitor to sign up for your email list via your opt-in form.

If you can accomplish all three of these, then you're already way ahead of the game.

If you're just getting started, the next few pages are going to rock.

If you already have an opt-in form and landing page and feel pretty comfortable with what you have, feel free to skip ahead to "deliver your lead magnet".

If you want to take the simple route and use your email service provider for your opt-in forms and landing pages, stick with me here.

Let's get started.

Choosing an Email Service Provider

Now that you know what it takes to create a landing page, you're going to need an email service provider to finish the process.

I recommend email service providers like MailerLite and ConvertKit, and here is where I'll start digging deep into how you can use them and how to set them up.

Why do you need an email service provider you might be wondering...

#1: Professional email capabilities

Well, while you can send out bulk emails to people via your Gmail account now (and you've gotten away with it so far), when you start to try and do that on a larger scale, you need to work with an email service provider or **face getting your Gmail account blacklisted and closed forever!**

Don't have a Gmail account but use something like Outlook or Hotmail? Same deal applies. You need a dedicated email platform with a domain branded email to manage your business leads.

There are a couple of great things that an email platform provider is going to do for you.

The most obvious is they will give you a way to collect (and store) your email subscribers. They will also allow your bulk emails to be sent with a higher chance of deliverability.

From there, you will be able to set up your automated email series and your regular newsletter.

#2: Easy landing pages

The main thing they will help you achieve are evergreen sales funnels that work 24/7 in bringing the right leads into your business and automating your sales.

This starts with your landing page.

But first, let's cover the basics.

There are quite a few email platform providers out there.

And I've used a good number of them... from Aweber to MailChimp to ActiveCampaign, MailerLite and ConvertKit.

If you and I were sitting down together having a coffee and cinnamon roll (yum!) and you said to me, "Hey Lise, what's your fav email service provider?" I'd tell you it's ConvertKit.

ConvertKit is free (with limitations) as well as providers like MailerLite and FloDesk.

But if you want to set yourself up for success right from the start, then ConvertKit is your best option.

Why?

A few reasons:

1. They have amazing support (who are responsive and helpful within the hour).

2. They have beautiful landing pages built-in.

3. You can create visual automations and funnels all in one screen.

One last thing, the most important thing to consider (beyond your needs) is your personal preference in using a provider.

Don't spend hours agonizing over which one you should pick. Go with the one you like the best.

The good news is, most of these services have a free trial period where you can take them for a test run and see how you like them.

Ready to start?

BEGINNER

Get started with simple email platform providers.

A couple of the more common ones that I recommend[1] are: MailerLite, ConvertKit - my personal recommendation (and what I use), and FloDesk.

Each of these are either free (ConvertKit and MailerLite) to start or have a low monthly subscription fee initially so none of them will break the bank to get started.

For most creative business owners who are just getting started, ConvertKit is a great option. It's very simple to use and has a clean interface.

You can figure out pretty much everything on your own right away. But, they also have a great set of help documents if you get stuck.

I recommend ConvertKit because it offers you quite a bit of flexibility in building your list. You can create your own opt-in forms, which you can then link to your email list and then schedule automated sequences from there.

They also have a number of really nice preset templates you can use for your landing pages, newsletter, and emails.

All you have to do is simply fill in the blanks and you are good to go!

You can learn much more about how to use ConvertKit to create opt-in forms and landing pages later on in this chapter.

INTERMEDIATE

For creative business owners who want a little more power and advanced techniques when building their sales funnels, consider options like OptimizePress, Instapage, and LeadPages.

For creative business owners who already have their own website and are more familiar with WordPress, there are also a number of plugins available such as:

- OptinMonster

- OptinForms

- Hellobar

- Sumo

These will give you more than just a vehicle to collect emails. They are actually pretty good "all in one" systems for building a sales funnel.

Here's one created with OptimizePress and OptinForms:

You can use these to create a professional opt-in and landing page in just a few minutes, as well as add in your thank you/download page into the mix. Many, like OptimizePress, are simply drag and drop.

They will do most of the hard work for you on the backend, leaving you to just plug in your bits and pieces.

The good thing about one of these options is you can use them to grow with your business over time, but they aren't too complex to be completely overwhelming from the start.

ADVANCED

For the advanced creative business owner who is looking to build a solid business, automate their income, and is interested in tracking and testing all their traffic... then you want to take a look at some of the more advanced options.

I like using OptimizeFunnels, Kajabi or ConvertKit for this. These are going to be a bit more complex and have a learning curve to get started, but the benefit is that they are quite powerful and give you a lot of options.

The main feature of using more advanced tools like OptimizeFunnels is they offer an all-in-one process that will take care of building your funnels, opt-ins, and sales processes.

You can build pages that look great and make traveling through your sales funnel easier over the long run.

With a tool like this, you can create a number of landing and opt-in pages. This allows you to offer different incentives for people coming from specific sites and then see which are generating more buzz and converting better.

Here's what a funnel looks like in OptimizeFunnels:

But setting this up is nothing you need to worry about right now, unless you're ready for this phase.

If you're in the beginner phase, all you need to worry about are the most basic functions.

You need an email platform provider that will cover the following:

- The ability to collect and store email addresses into a list

- The ability to send out an autoresponder series

- The ability to send out a regular newsletter

- The ability to create landing pages quickly

That's pretty much all you need to start.

Any good email service provider is going to offer you that and a lot more.

Some of the other things you (ideally) want your email service provider to include is the ability to use a plugin that will work with your website to collect the emails, and have a landing page builder built-in.

To start, you don't necessarily need that stuff, but they will be nice to have as you continue to build your list and grow your numbers.

Why ConvertKit is Right For You

For beginners or creatives wanting to keep things simple, the email service provider I'd recommend using is *ConvertKit*.

There are a couple of reasons why I like ConvertKit. One big bonus is that it's free forever with limitations.

Here's a quick snapshot of their pricing tiers based on if you had 1,000 subscribers or less:

How many email subscribers do you have?

0 ——●—— 500k+

1,000

Monthly | Yearly

(Get two months free)

	RECOMMENDED	
Free	**Creator**	**Creator Pro**
For new creators building their list	For growing creators automating their work	For established creators scaling their business
$0 /month	**$29** /month	**$59** /month
Free for up to 1,000 subscribers	For up to 1,000 subscribers	For up to 1,000 subscribers
Sign up free	Start free trial	Start free trial
Unlimited landing pages	Unlimited landing pages	Unlimited landing pages
Unlimited forms	Unlimited forms	Unlimited forms
Unlimited broadcasts	Unlimited broadcasts	Unlimited broadcasts
Audience tagging and segmentation	Audience tagging and segmentation	Audience tagging and segmentation
Sell digital products & subscriptions	Sell digital products & subscriptions	Sell digital products & subscriptions
Community support	Live chat & email support	Priority live chat & email support
	Free recommendations	Free recommendations
	Paid recommendations	Paid recommendations
	Free migration from another tool	Free migration from another tool
	Automated email sequences	Automated email sequences
	Visual automation builders	Visual automation builders
	Third-party integrations	Third-party integrations
	One additional team member	Unlimited team members
		Newsletter referral system
		Subscriber scoring
		Advanced reporting

So for the vast majority of creative business owners who are getting started, ConvertKit is a very easy choice, even if you want to start on the Creator plan *(which I recommend so that you can send automated email sequences).*

ConvertKit has well over 30 different templates you can use on your site for opt-ins, or you can make your own if you want.

They offer a WordPress plugin as well as integrate with most websites, which allows you to easily capture email subscribers with just a few clicks from within your website.[2]

If you're not able to use any of the integrations, you can easily embed the code on any site as well.

Of course, ConvertKit also allows you to set up a number of subscriber tags, automated email sequences, and subscriber segments as well. So they cover all of your basic needs plus a little more.

Getting Started with ConvertKit

Below I'm going to dive in with some basic tips on getting started. For more detailed information, check out the ConvertKit Overview here: https://convertkit.com/training.

Once you've signed up for a ConvertKit account, the first thing you want to do is create a tag.

To do this, follow these steps:

1. Go to the ConvertKit subscriber area,

2. On the right-hand side, you'll see the option to "create a tag".

3. Create a tag for your first lead magnet. Something like "LM: Income Projection Calculator" for example.

In the example below, you'll see a number of tags already created and the number of subscribers associated with that tag.

⬚ Tags

1 / 3 - Researcher/Experimenter
(Investigator/Martyr) 40 SUBSCRIBERS

1 / 4 - Researcher/Community Builder
(Investigator/Opportunist) 8 SUBSCRIBERS

1. Get ideal clients to join email list 3 SUBSCRIBERS

2 / 4 - Natural Genius/Community Builder
(Hermit/Opportunist) 52 SUBSCRIBERS

2 / 5 - Natural Genius/Teacher (Hermit/Heretic) 7
SUBSCRIBERS

2. How to sell to email subs 0 SUBSCRIBERS

3 / 5 - Experimenter/Teacher (Martyr/Heretic) 56
SUBSCRIBERS

3 / 6 - Experimenter/Role Model (Martyr/Role
Model) 6 SUBSCRIBERS

3. How to optimize content to sell 24/7 1
SUBSCRIBER

4 / 1 - Community Builder/Researcher
(Opportunist/Investigator) 9 SUBSCRIBERS

4 / 6 - Community Builder/Role Model
(Opportunist/Role Model) 45 SUBSCRIBERS

5 / 1 - Teacher/Researcher (Heretic/Investigator)
40 SUBSCRIBERS

5 / 2 - Teacher/Natural Genius (Heretic/Hermit) 8
SUBSCRIBERS

6 / 2 - Role Model/Natural Genius (Role
Model/Hermit) 47 SUBSCRIBERS

6 / 3 - Role Model/Experimenter (Role
Model/Martyr) 6 SUBSCRIBERS

A - Live Workshops 0 SUBSCRIBERS

AAB Participant June 2023 15 SUBSCRIBERS

AAB Participant Sept 2023 19 SUBSCRIBERS

AAB Waitlist Jun 23 10 SUBSCRIBERS

LM: GYB affirmation cards 8 SUBSCRIBERS

LM: H&G Vault 234 SUBSCRIBERS

LM: HD Custom Biz Plan Blog Post Sub 16
SUBSCRIBERS

LM: HD Custom Biz Plan Intro FB Lead Gen Ad 526
SUBSCRIBERS

LM: HD Custom Biz Plan LGA FB 340 SUBSCRIBERS

LM: KYH Toolkit Access 45 SUBSCRIBERS

LM: L180 Guest Post Author list 1 SUBSCRIBER

LM: L180 Guest Post Upwork 1 SUBSCRIBER

LM: Lise Cartwright Author Subscribers 25
SUBSCRIBERS

LM: MBD as Freebie 340 SUBSCRIBERS

LM: MMMP Lizzy's Xmas Party 2022 148
SUBSCRIBERS

LM: NGN Book Subscribers - Action Guide 14
SUBSCRIBERS

LM: OFS Guide Book Freebies 0 SUBSCRIBERS

LM: Permafree Book (Pimp Your Profile) 1
SUBSCRIBER

LM: SH 7-Day Challenge 35 SUBSCRIBERS

LM: SHB Free PDF 16 SUBSCRIBERS

LM: Side Hustle Blueprint book 1 freebies 19
SUBSCRIBERS

LM: Side Hustle Blueprint Book 2 freebies 52
SUBSCRIBERS

This is just a quick snapshot from my own CK account... I have hundreds of tags.

Just a note, if you already have a list of people in a spreadsheet or from another platform, you can easily import, or copy and paste the names into your account.

Remember, only include the emails of people who have specifically given permission (aka have confirmed they have opted in) for you to contact them, no spamming allowed.

Next, create your opt-in form.

Go to your ConvertKit menu, click on Grow > Landing Pages & Forms..

Click "Landing Pages & Forms".

Then click on the button "+ Create new" - the red button in the bottom of the window.

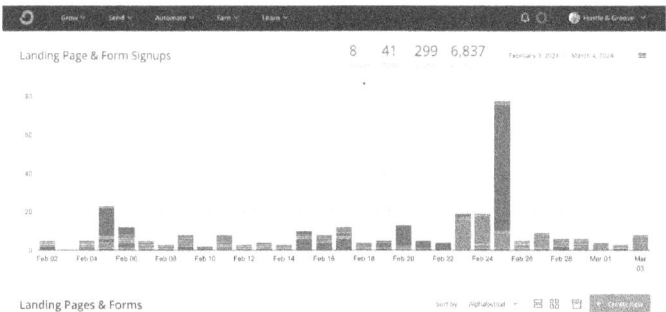

Then click on what you want to create - either a form (for embedding on your website) or a landing page for use as a standalone page or integrated to your website.

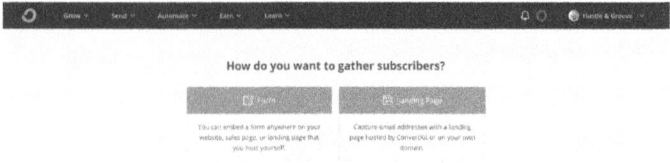

Let's create a landing page. Choose that from the options provided and then you'll be presented with templates to choose from (this is just a quick snapshot, there are loads of options).

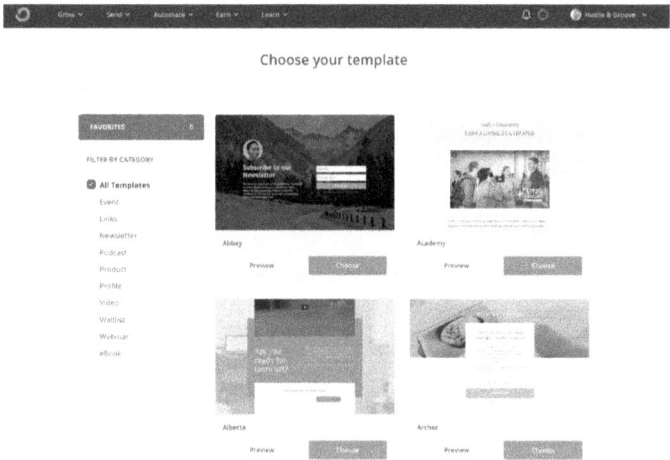

Pick from one of the design options and start filling in the contents. You'll notice an area over on the right-hand side. This is where you can change colors, fonts, and images.

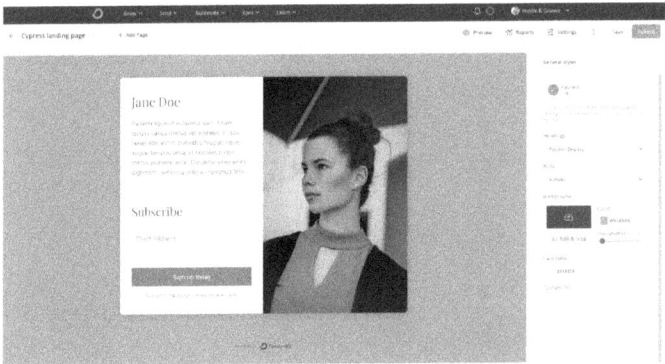

Don't forget to name your landing page before you go too much further. You can do that by clicking on the little pencil at the top of the page, just below the ConvertKit menu bar.

ConvertKit has great training videos and live training workshops on every step of this process, so make sure you check them out in their video tutorials knowledge base here: https://convertkit.com/training.

If you don't have a website you should keep in mind that ConvertKit gives the option of giving you an actual link to the landing page that you can also use to promote your lead magnet with.

Once you've got your design set up, click Save and Publish.

Awesome. Now you've created a tag and have your opt-in form/landing page set up!

Congratulations. That wasn't so hard now, was it?!

Now, let's focus on getting your lead magnet delivered to your new subscriber.

In the next section, I'll discuss what that looks like and get you started with your first email to your brand new email subscriber.

---

1. w w w . h u s t l e a n d g r o o v e . c o m / m a i l e r l i t e, www.hustleandgroove.com/convertkitemail, and www.flodesk.com. **Please note** that some of the links included inside this book are affiliate links. This means that I receive a commission should you choose to purchase using my link. You are under no obligation to use this link, nor does it cost you any more to use my link. I only recommend what I personally use.

2. For a full list of ConvertKit integrations, check out this page: https://convertkit.com/integrations

Step 3:

Deliver Your Lead Magnet

Now that you've got your lead magnet created and your landing page set up, it's time to make sure that your brand new lead gets what they signed up for.

So how do you deliver your lead magnet?

You have two options:

1. Via email (as a link or direct download)

2. On a specific download page

Those are the only two options.

Pretty simple.

If you want to keep it simple, I'd recommend setting this up to deliver via email, which is very easy to do.

There are a couple of reasons for this.

The first is that you have an opportunity to start connecting with your ideal person through your initial lead magnet delivery email.

You want them to *want* to open your emails, so you start "training" them from the beginning to associate your name with an email they want to open.

Most email platform providers will give you the option to deliver your lead magnet within the confirmation email. Either as a link or something you upload.

Now, if you already have a website and want to deliver the free gift via a download page, that's perfectly fine too.

Here's what that could look like:

The easy part about using a download page is that many of the tools already have a built-in download/thank you page template that can easily be tied in with your main landing page.

When you're setting up your landing page, simply add a Thank You or Download page to your opt-in flow.

Then once someone opts into your mailing list, they will be directed to a Download or Thank You page, and on that page, they will be able to access the lead magnet.

Both methods work well for delivery, so the choice is up to you!

Here's an example of what your lead magnet delivery email looks like when using ConvertKit:

Just one more step...

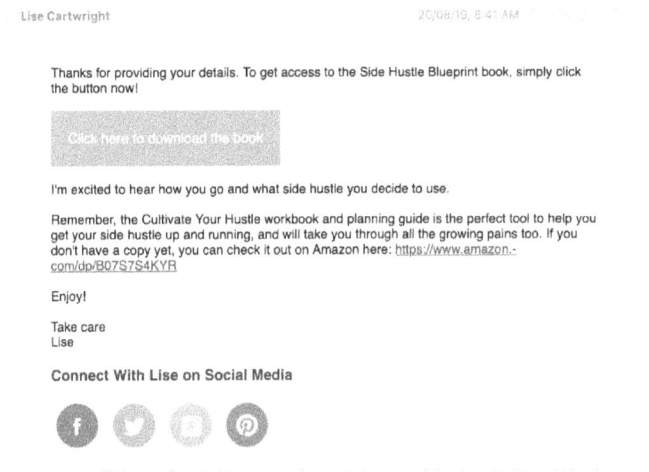

Lise Cartwright   20/08/19, 6:41 AM

Thanks for providing your details. To get access to the Side Hustle Blueprint book, simply click the button now!

Click here to download the book

I'm excited to hear how you go and what side hustle you decide to use.

Remember, the Cultivate Your Hustle workbook and planning guide is the perfect tool to help you get your side hustle up and running, and will take you through all the growing pains too. If you don't have a copy yet, you can check it out on Amazon here: https://www.amazon.com/dp/B07S7S4KYR

Enjoy!

Take care
Lise

Connect With Lise on Social Media

And that is the end of **Part 1: Amplify Your Reach.**

Your Key Takeaways:

- How to identify your best strategic freebie

- How to capture your ideal people with opt-in forms and landing pages

- How to deliver your lead magnets

In the next two sections, we're going to dive into converting your new email subscribers into customers and what a simple sales funnel looks like.

And don't freak out! The sales funnel piece is a lot easier than you're thinking... so let's jump straight to it.

PART 2: ENGAGE

"Why waste a sentence saying nothing?" **~ Seth Godin**

In this second part of the book, we're going to look at how to convert your new subscribers into raving fans then into loyal customers.

What makes them stop and say, "Yes! I want this now!" What makes YOU do this with your fav products or services?

As I mentioned previously, everything that I'm sharing with you is neither 'black hat' or icky. It's about providing value at every point of the customer journey.

You should treat your new subscribers with the respect and kindness they deserve.

So under no circumstances are you trying to sell them products or services that don't genuinely help them.

Because ain't nobody got time for that!

So what am I talking about here? The first step in conversion is nurturing your awesome subscribers and then converting them into your best friends and paying customers/clients.

I call this part of the process "engage".

If you get this right, you'll start to automate your income generation and create a life and business you truly love.

Let's get started!

Step 1:

Welcome Sequence

Now that you've got your lead magnet, landing pages, and lead magnet delivery email set up, let's make sure your new subscriber feels welcome!

We want them to stick around for a while, right?

We want them to be so in love with you and all that you provide, they are literally clamoring to buy your thing.

The best way to make them feel welcome and loved right from the start is to create a great welcome sequence that indoctrinates them into your community.

Now we get to the fun stuff. Setting up your initial welcome sequence for your new email subscribers.

But these aren't just any email subscribers. No. These are your ideal leads. The people you want to work with the most.

So where does the welcome sequence fit in?

Basically, after your lead magnet is delivered, your welcome sequence begins.

You've done the work to get them on your email list, now we need to convert them into the right part of your business that best meets their needs.

The way to do that is to make sure that you start things off on the right foot with a really good welcome sequence.

What is a welcome sequence?

It's a series of automated emails that go out to your new subscriber based on an event set up in your email platform. Generally, it's triggered to send out 1 day after joining your email list.

It should educate them about you and your business as well as provide them with more value and opportunities to learn more about you and what you provide.

We'll also start to *present the opportunity* for them to buy from you too.

Ideally, your welcome sequence will consist of 2-5 emails, all scheduled to go out one day apart.

Wondering what your email sequence should look like? Let's create one together now.

Let's follow a fictitious business owner, Jane, as she creates her welcome sequence. Her business centers around teaching time management and includes books and a coaching program.

What follows are examples of the emails that would appear after I signed up to Jane's list.

The subscriber's name would, ideally, be part of the information you've captured in the opt-in form and can be inserted using code. That way, each email is personalized to the subscriber.

Check the help section of your email provider for specifics as well as the resources page link for a quick tutorial.

Day 1 - Email 1: Strategic Freebie Delivery

The first email is the freebie delivery email. This is the email that will immediately be sent to anyone who subscribes to your list or opts-in for your lead magnet.

This email is going to do a couple of things.

First, and most importantly, it delivers your free thing.

Since your leads signed up for your list in exchange for something, you absolutely want to be sure they get that something as quickly as possible.

The second thing that this email does is confirm their email address. Important if you want to ensure you're meeting GDPR regulations and CAN-Spam Act laws.

For more information about what you need to do in regards to these regulations and laws, I recommend reading   up   on   this   here:

https://segment.com/growth-center/email-marketing/rules-regulations/ .

Note: If you've set up your incentive email (in ConvertKit), this is the email you would send.

Here's a good template you can use to get you started:

Subject: Just one more step...

Hey Lise,

Thanks for providing your details. To get access to the 90-Day Plan simply click the button now!

[Click here to download the plan]

I'm excited to hear how you go and what your next 90 days look like!

Remember, the Cultivate Your Hustle workbook and planning guide is the perfect tool to help you get your side hustle up and running, and will help you through all the growing pains too. You can apply your 90-day plan to your business easily once you've finished reading the workbook.

If you don't have a copy yet, you can check it out on Amazon here:

Enjoy!

Take care

Jane

PS: *Keep an eye on your inbox for even more value coming your way over the next few days.*

You can see here that this email is pretty simple. It delivers the lead magnet, it shows a little bit of your personality and it lays out what the reader can expect from you.

It doesn't have to be too long to get the point across.

Day 2 – Email 2: Welcome email (the start of your official welcome sequence)

This is going to be the first email in your welcome sequence.

This email is where you want to introduce who you are, what your business is about and how you help people.

It's where your personality will shine.

Here's an example:

Subject: Uh, are we gonna gel?

Hey Lise,

Now, I just wanna take a minute and say how honoured I am that you joined the H&G community.

But before I tell you all my secrets and share TMI, I kinda need to know if we're gonna gel.

Because if you're not the type of person I can help, or if you don't resonate with what I say, then I don't wanna take up space in your sacred inbox. Truly.

A little about me:

Over the past 10+ years, I've been helping awesome people such as yourself, figure out how to crush their goals and increase their productivity without working more hours.

I like to think of myself as your virtual biz guide.

One of my readers once described me as "the friend who pulls no punches and tells you what you don't wanna hear with a genuine, heartfelt smile."

I'd also add that I'm an awesome karaoke singer (catch me at shower time!), I use GIFs and emojis all the time, and I make a mean mac and cheese.

Who I Can Help Most, and Who I Can't:

There are a definite group of peeps I tend to get awesome results for Lise,... and there are also others who just aren't the right fit for me and the community.

Let me tell you about each camp real quick so that you and I are on the same page...

I get awesome results for:

- Authors, coaches, and creative entrepreneurs who want to leverage their time to implement systems in their online business, who are willing to take responsibility for taking action, ie, you're willing to do some work to see the magic happen.

- Authors, coaches, and creative entrepreneurs who want to grow and scale their online *income and are looking for help with mindset optimization and simple systems to automate their income.*

On the other hand, I generally won't be able to help people who...

- *Spend time on the Internet just to make fun of others and leave comments like, "you suck." I only use encouraging words, you should do the same.*

- *Are racist against any race or culture. I don't tolerate these type of beliefs, you should feel the same.*

- *Aren't willing to put effort into taking action. The stuff I share with you won't come to life unless you actually do the work based on what works for you.*

If you've read this far, I'm gonna assume you gel with me and fit right into that first group, Lise.

Now we can continue with the same goals in mind.

But, I also know I'm not a good fit for everyone, so if you'd like to "peace out" now, no hard feelings at all.

Go ahead and hit the "unsubscribe" link at the end of this email. I wish you all the best.

Coolies?

Still here? Awesome. We gel just fine!

Tomorrow, I'm sharing some of my fav resources with you.

I'll see you then.

Take care

Jane xoxo

If you want to increase engagement, ask your new subscribers a simple question that they can click reply and answer to.

Day 3 – Email 3: More resources

This email is designed to help you showcase your expertise while also providing your new subscribers with resources... free or paid.

There is no right or wrong amount of resources to provide.

Think about what makes the most sense for your new subscriber.

If you decide to provide all free resources, make sure they are optimized with a call to action that is meaningful to you and your business goals.

Let's take a look at an example.

Subject: Only the best for you (my top three resources)

Hey Lise,

Was just thinking of you today and thought you'd like the inside scoop to my 3 most popular content pieces.

I hope you enjoy your free gift too. Have you cracked it open yet? Started implementing...

You'll find actionable tips in each of these resources, based on where you're at in your business.

Click one (or all) to dive right in!

- **How to create a simple sales system** (This is how you create a an automated funnel using one simple tool)

- **3 Easy Steps to Start a Digital Product Business** (If you're just starting out, follow these steps to get yourself making money in the next 30 days!)

- **Explode Your Productivity With a 90-Day Plan** (I've been using 90-day plans for over 5 years now and they not only make you productive, but provide you with flexibility to change course when needed!)

Enjoy.

So excited to connect with you further.

Take care

Jane xoxo

PS: I'm so honored you've granted me sacred access to your inbox.

To ensure you don't miss out on any content or updates, make sure you whitelist **jane@handg.com** in your inbox! I'd hate to see our quick tip juicy carrots go unnoticed in your spam folder.

This is one of my favorite emails in the welcome sequence.

It does a great job of providing more help to your new subscriber, while also providing your subscriber with an inkling of just how much support you can provide them with.

Any of these resources could link to a paid offer if it made sense to do so.

If you were planning to offer a paid offer, make sure it's under $20. You'll have a higher conversion and it provides your new subscriber with an idea of how you provide support.

Thus, the *know, like, and trust* journey has begun!

Day 4 – Email 4: More of your story + a soft sell

Inside this email, you'll want to share more of your story.

This is where you'll want to start connect with your audience in a way that shows that you have had a similar experience to what they might be in right now.

You'll need to have a pretty good understanding of your audience and where they are at.

Let's look at an example.

Subject: My #1 time management strategy

Hey Lise,

Back in 2019, when I started my business, I remember thinking that I'd be writing books full-time forever.

It was the ultimate dream... to write a book, launch it, and watch the money (passively) roll in.

And while I've been blessed for that to be (mostly) the case, over the last few years, I felt the need to do more.

To share more.

And one of the things that my readers and subscribers kept asking me about was my systems.

They would send me emails saying things like; "Lise, how are you able to do so much? Do you have a secret time machine?!"

I hadn't realized it, but my readers and subscribers had pointed me in the direction of my new products... courses and coaching around systems and productivity.

If you think it sounds simple, that's because it is.

When I first started trying to master time management, I constantly shot myself in the foot by trying to take on too much.

You know how it goes, right?

You want to fix it all so badly, you'll do whatever it takes. And it seems to work, at least for a few days.

Until...

...it's so overwhelming that it actually starts causing more havoc in your life.

Then you're right back where you started.

I did that more times that I want to even admit.

Finally, I started with the tiniest of baby steps.

That helped me build up the confidence to move forward, and became the start of my 5 Point Time Management system.

Now I share this system with all of my coaching clients. In fact it's the very first thing we cover after they get inside the program, that's how much it helps.

As a special bonus gift for you, I'm giving you a secret link to the discounted version.

[link to 5 Point course]

Hopefully, you will be able to use this "secret weapon" to get started on your own path to being better with time management and productivity

Talk soon,

Jane

P.S. Here's the link again to my 5 Point Time Management course [link].

This email helps to accomplish a couple of things.

First, it shows that you understand where a large portion of your audience is at. A quick story to set the stage is always a good thing.

Second, it does a great job of leading your new subscriber to a logical next step... which happens to be a paid offer.

When you talk about how things work for your clients or students that's a subtle hint that you have been able to help other people get success.

And it tells them you have something to sell. Subtly, but enough that it's not a surprise...

This will lead into the final email in this welcome sequence.

Let's take a look.

Day 5 - Email 5: Closing the loop

In this final email in the welcome sequence, we're going to close any loops that we may have left open in our other emails.

In this particular email, we're going to share another story about how the business got started... in the form of a video.

We're also going to remind them about the offer we made in the previous email.

Let's take a look:

Subject: *How it all began...*

Hey Lise,

I thought it'd be pretty cool to share my story with you in one short snackable video about how I got started and created the H&G brand.

Click here to see the story.

So take heart on your entrepreneurial journey.

You can do this!

Don't for a second doubt yourself.

You got this and I got your back, always.

See you in the next daily email.

And don't hesitate to reach out if you need help. I'm here to support you in any way I can.

Take care

Jane xoxo

PS: Did you get a chance to check out The 5-Point Time Management course! It's a simple system to help you claim back your time and increase your productivity... without doing more. Here's your last chance to access the discount link [link].

So here you should be able to see the power of the soft sell in the P.S.

There are several ways that you can write this, especially in the emails you send after your welcome sequence.

My personal favorite is something along these lines:

PS: When you're ready, here's how I can support you right now (enter the name of your offer and the benefit and provide a link to it).

RESOURCE: You can grab the Welcome Sequence Template from the resources page here: www.hustleandgroove.com/cyaresources

By now you have given quite a bit to your new subscribers.

You've respected their time and space, not spammed them, offered them more free stuff, all of it valuable. Then you simply present them with an opportunity to buy from you.

Most people feel squeamish at this part. That's totally normal. But, if you don't ask, it's very hard to get people to buy. Remember that!

So what happens at the end of this welcome sequence?

What happens if someone buys and if they don't?

That's what we're going to cover next!

But first, let's get this set up in ConvertKit.

SET UP WELCOME SEQUENCE IN CONVERTKIT

It's time to create a sequence in ConvertKit (you will need the paid version to do this), the first step in getting a welcome sequence live.

From the domain menu click Send > Sequences.

Click on "+ New Sequence" to get started.

Next you'll need to choose a template for your sequence. If you haven't yet set these up, you'll be prompted to do so.

Otherwise, simply click on a template to open the sequence editor.

Choose a template for your sequence

HG Newsletter Template

Weekly Email Template

HG Newsletter New

HG Sequence Template

Coach

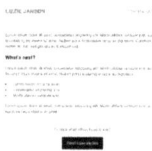

Copy of Copy of Creator

Next, you'll be taken to the editor where you can write your email sequence. Don't forget to name your sequence first (in the top, left-hand corner where there is a little pencil).

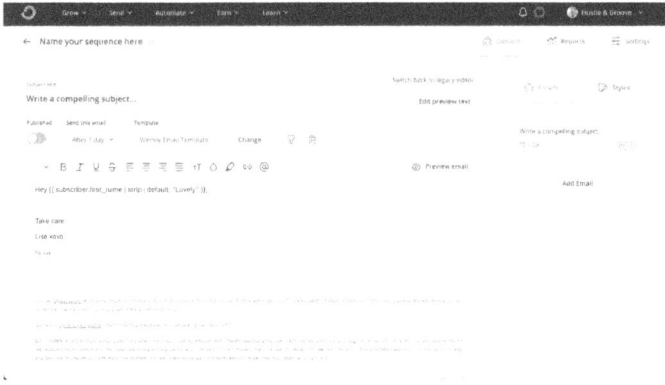

From here, you'll want to write your subject line and the email for the first email (following on after the incentive email delivers your freebie).

Once you've written the first email, click on "Add Email" on the right-hand side to add the next email.

Write a subject line and the body of the next email. Rinse and repeat until the entire sequence has been built out.

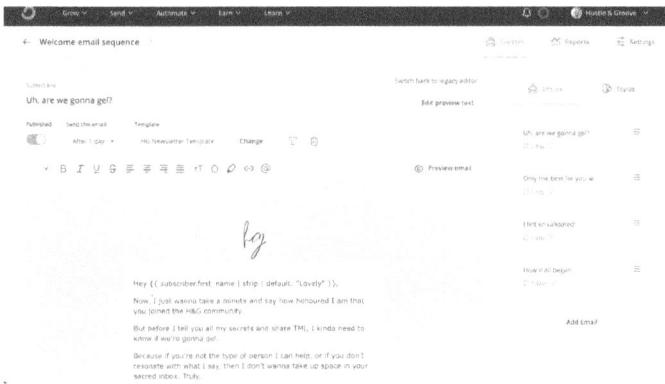

You can add images, headings, text blocks. But remember, keep it simple.

When everything looks good, it's time to set up the automated part of the sequence.

You can do this in two ways:

1. Create a rule that when someone joins a form (or landing page) they get added to this sequence

2. Create an automation (you'd only do this if you wanted multiple things to take place, such as adding a tag, starting another sequence based on a link they click etc)

I'd recommend that you simply create the rule.

To do that, you click on Automate > Rules.

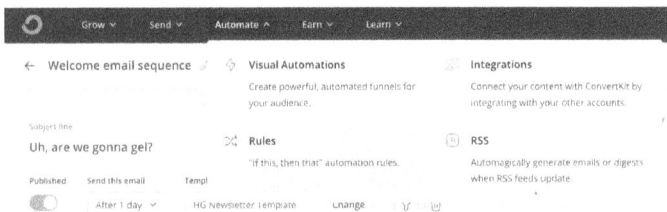

On the next screen, click on the red button "+ New Rule".

You'll see this screen:

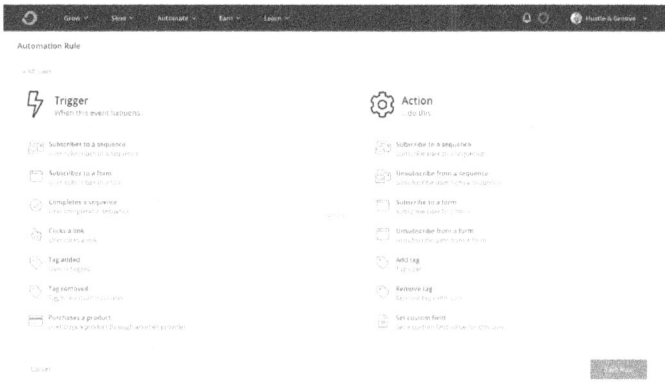

The first step is to choose the Trigger - Subscribers to a form.

Once you click that, you'll be able to choose the form for your lead magnet.

On the Action side, you'll choose "Subscribe to a sequence" and then choose the sequence you just created.

Now you're ready to go!

Everytime you create a new form or landing page, simply repeat the process. This process is similar in other platforms too.

TIP: You do not need to create a new sequence each time. You can simply use the same sequence.

Now that you've got your welcome sequence set up, let's jump into the main part of this section of the book... converting your new subscribers into customers (if they aren't already)!

Step 2:

Nurturing Your Leads

This is a crucial step where most business owners end up making a mistake. They take the time to set up an automated welcome sequence and send it out, but then... crickets.

They don't do anything after that. It's a big problem, and I guarantee it will help you lose subscribers (and future customers).

That is the complete opposite of what we're trying to accomplish.

So, when it comes to your email list, even after your welcome sequence is written and sent, your work isn't done.

In this next chapter, we'll look at a couple of the most common options that follow the welcome sequence.

Let's get started.

SENDING NEWSLETTERS

One of the more common things to set up after your welcome sequence is finished is a regular newsletter.

A newsletter (also called a broadcast email inside ConvertKit) is a message that you write and send out at the same time each day, week or month.

You're probably familiar with these in your own inbox, daily or weekly updates from your local grocery store or monthly updates from your favorite sports team; both are categorized as newsletters.

With the privacy and spam changes that came into effect on February 1, 2024, it's now even more important that you email your subscribers regularly.

If you don't know what those changes are or what they mean, here's a quick recap:

Email service providers like Gmail, Yahoo, AOL, and Microsoft have cracked down on privacy and spam issues.

If you're sending emails to subscribers, these are the issues you'll need to ensure you've addressed:

1. *Configure email authentication with DKIM, SPF, and DMARC. You will need to do this at your domain level (Aka your branded business email. Mine is lise@hustleandgroove.com) and at your email marketing platform level (ConvertKit, MailerLite etc). You can learn more on the ConvertKit website.[1]*

2. *Do not impersonate Gmail "From:" headers.*

This means that you need to use a domain verified email address to send emails from (like mine – *lise@hustleandgroove.com*).

3. Maintain a spam rate below 0.3% and send relevant emails. This is something you'll need to monitor each month.

4. Provide an easy one-click unsubscribe system and process unsubscribes within 2 days.

Most business owners should look at sending out newsletters either twice a month or weekly, depending on how much valuable information you can share. This strikes a nice balance.

Also, be sure to let your email subscribers know how often you will be sending them regular emails. Consistency is very important here to keep your deliverability health good, open rates high and click-through rates even higher.

Your newsletter or regular emails can be about what's happening in your life, valuable content for your subscribers, a new blog post, YouTube video or podcast episode to share.

RESOURCE: 60 Newsletter Prompts + Storybank Template (available inside the Resources Hub)

Newsletters don't have to be long or drawn out, you can easily take about 10 to 30 minutes to write one and click send. They are a great way to keep in touch with your audience.

Here's an example of a newsletter to my H&G community (what I call my email subscribers):

(you can see it live here: https://ckarchive.com/b/4zuvheh589vxqt6ovveol a3drwv77)

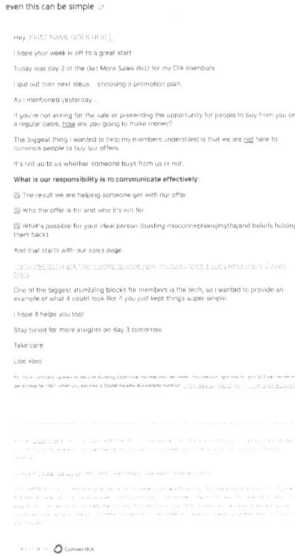

DECIDING ON NEWSLETTER CONTENT

One of the things I get asked the most by my List Building Collective members is, *"how do you come up with what to write in your emails Lise?"*

It starts with my 'storybank'.

The storybank is a document where I keep all my stories about:

- *How I started my business*

- *Why I started my business*

- *What my journey has been like*

- *People who have impacted my business*

- *Any struggles I experienced along the way*

- *The little wins and big wins along the way*

This document is something that I refer to a lot, not just when it comes to writing emails.

I also use it to help with content for my sales pages and sales emails.

But the storybank is just the start.

As someone who writes emails every day, I need ongoing stories and content to share with my audience.

There are two other things that I do to be able to write daily emails with ease.

#1: Daily story prompt

This is very simple and something you can implement yourself today.

On my phone, I have a daily reminder set (for 5pm) that asks the following question:

What happened today? Capture your stories.

5:58   .ull 📶 🔋

✕   ✏️   •••

DEFAULT LIST

What happened today? Capture your stories

Friday, 15 Mar at 5 pm

Repeats daily

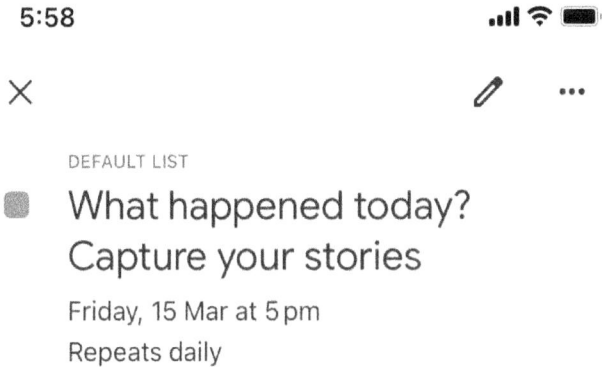

This serves as a prompt for me to do exactly what it says... capture stories.

At this point, I sit down or go for a walk and speak about my day.

I use Otter.ai (the app) to do this. It's a great tool for taking your voice messages and turning them into text at the same time.

When I do this, I simply recount my day.

The process of doing this will generally provide me with some stories that I can share with my audience.

When I sit down to write my daily emails, the other thing that makes this easy is...

#2: Being clear on my main message

Some might refer to this as 'brand messaging'.

I refer to this as the phrases I repeat again and again.

Or the ways I want to help my audience.

It starts with this question: *what am I helping my audience with?*

Why do you do what you do? This goes beyond your personal reasons for starting a business.

At the core of your business, I believe that you truly want to help people in some way.

For me I'm here to empower and inspire my audience to believe that their businesses can be easy and fun.

That there are no one-size-fits-all strategies when it comes to running their business.

To be more you.

To focus on the marketing activities and offers that feel easy and fun for you.

When I sit down to write my daily emails to my H&G community, it starts with asking myself this one question:

How can I empower and inspire my community to believe that their business can be easy and fun, that they can do the things they want to do in their business, even if it goes against what others are saying works?

I then pull out my phone, scroll through my Otter.ai app and look for stories that I can share that support this message.

Here's an example of an email that does this:

(you can see the live example here: https://ckarchive.com/b/8kuqhoh0d6edvb3n66m nqfzpzg399) "you changed your mind?!"

Hey [FIRST NAME GOES HERE],

I hope you had a great weekend.

Mine was spent reminiscing... tomorrow (Feb 20th on this side of the world) is our 10th wedding anniversary.

As someone who travelled in her late 20s, I have a number of great friends located on the other side of the world.

We were blessed to have three of them join us on our wedding day and be part of our bridal party.

I think it's one of the only things that I truly enjoy about FB these days... the photo memories that pop up!

In any case, I hope you had a great time doing things you love. 😊

The other thing I did this past weekend was review my upcoming book, Cultivate You!

Back in November 2023, I wrote this book in 14 days.

Since then, I've let it settle and marinate... and I'm glad I did.

That book has now turned into three books! 📚

One main book and two workbooks.

My hubby walked into my office yesterday and said, "you changed your mind?!"

Yes. Yes, I did.

And here's the thing about that...

It's my book. My business.

If I want to change my mind about how I present the content, even after I've had a cover designed and beta readers' feedback, I can.

The biggest reason for changing my mind?

Paying attention to my audience.

Seeing where they get stuck.

I have a front-row seat to this inside the Digital Income Accelerator membership.

I can't wait to release these three books in March!

Stay tuned for how you can be part of the exclusive launch team and unlock special bonuses.

Remember: it's your business. If you want to change your mind about something and it FEELS good.

Do it. It's always your choice. Your decision.

Take care

Lisa xmin

P.s This is a note that as a Digital Income Accelerator member I am privy to inside stuff that other people don't change. This includes the Live Monthly Calls and membership. I am sure this also means are really, but as a free member you are only accessing a part. The trigger are already ways of being a member. [link to link]

P.p.s [illegible text line one] ...

P.p.p.s [illegible text line two] ...

[illegible text block of several lines] ...

Built with 🟠 ConvertKit

The key to making these emails relevant to your audience is in the last few sentences of the email.

80% of the email is the story.

To make the email relevant and practical for my community, I started with the word "Remember".

Then I went into what I want them to know and understand. This is where I empower them.

My story is designed to inspire them.

When you're crafting your own newsletters, keep this in mind.

NURTURE SEQUENCES

If you want to take things to the next level, then consider setting up automated sequences to handle the following situations:

- New customers

- People who don't buy your thing (retargeting campaigns)

This would mean setting up an automated sequence (automation in ConvertKit) based on an action your subscriber takes or doesn't take.

If someone buys your product or service, what emails would they get next?

You'd at least want to provide them with access to the offer, whether that's via an email with their login details in it or something else.

What else do they need? How will you continue to support them as they work with you or use your product?

And what about the people who don't buy right away? What happens to them?

Sure, you can simply continue to send them your newsletter, that's a good starting point. But what about if you sent them through an email series that nurtured them further around the solution you're providing?

What if the timing just wasn't right for them? How can you help them while they figure things out?

What type of emails might they need to receive during this period?

Personally, I like to retarget new subscribers with a quick 3-email sequence (that they receive before they start getting my daily emails) that gives them another opportunity to buy my thing.

You can see a template for that in the resources hub.

RESOURCE: Retargeting Sales Email Sequence Template

Bottom line: Take the time to think through these ideas and then map them out in your email platform provider.

By setting these pieces up, you'll go a long way to converting more subscribers into buyers and then retaining those buyers long-term.

Whew, that was a lot to cover in this part of the book, but you made it!

Your Key Takeaways:

1. You should know your email platform provider inside out

2. Created and implemented a welcome sequence

3. Decided on ways to nurture your new & existing subscribers

In the next part of the book, we're going to focus on *captivating* those people who bought, but also those who have yet to buy.

Jump in and get started.

---

1. Source: https://help.convertkit.com/en/articles/2502558-verify-your-domain-to-optimize-your-deliverability

.

PART 3: CAPTIVATE

"When you treat customers right, they come back... and they tell others." ~ **Shep Hyken**

In this third part of the book, we're going to look at how to captivate your new subscribers and customers so that they never want to leave.

Why should you care about captivating them?

Because it's a lot easier to keep a subscriber or customer than it is to go and find new ones... hello? We're not chasing the sale anymore, remember?! If you implement this phase, you'll literally be able to start predicting your income.

You should treat your subscribers, customers, and clients with the respect and kindness they deserve. Like they are your family.

Nurturing them should never stop.

What am I talking about here?

I'm talking about creating amazing experiences for your new subscribers and for your customers and clients that make them stick around.

If you get this part of the process right, you'll never have to chase the sale again.

You'll be able to relax and focus on serving rather than worrying about trying to get new customers or clients.

Consistent income is our goal. It's not our final destination — because it's not all about the money — but it does matter. And if you aren't consistently earning then you're focused on chasing the sale, which sends out negative energy to the Universe and your potential leads. We don't want that.

So the sooner you can consistently generate income, the sooner you can focus on growth and positive energy.

Life looks very different when you're not worried about when you're getting paid.

Let's get started!

Step 1:

The WOW Experience

If you really want to captivate your audience, you need to learn how to WOW your customers, clients, and email subscribers.

You need to create an amazing experience so that they never want to leave you.

This is about activating those amazing people that you want to work with or serve into action.

So what does that look like?

First, you need to really understand your target audience. Do you know what they are struggling with, what's keeping them up at night?

Do you know what the difference is between a successful creative business owner and one who's just getting by?

The **successful creative business owner** *does whatever it takes to get the answers she needs in order to serve her ideal customer or client. She is* *constantly checking in with her people, seeing how she can best serve them.*

The **"just getting by" creative business owner**, *on the other hand, says things like, "Oh, I'll get to that next week. This week, I'm focused on writing 3 new blog posts." She has no idea that her blog posts aren't even being read...*

It's time to get a little uncomfortable and do the things that put a catch in your breath.

Step out of your comfort zone.

Your business deserves this information and your target audience deserves your awesomeness.

Don't let a little thing like fear stop you from taking action.

And that first action you're taking? Surveying your audience! Asking them the questions you need to know in order to continue to serve them.

You can incorporate this into your welcome sequence or you can have it as a standalone email that you send out periodically to your audience.

Personally, I do both.

For my new subscribers, in my welcome sequence, I ask one simple question (this question is asked inside the third email in my welcome sequence):

"So... Let's start with you telling me *what you're looking to achieve in the next 90 days to grow your business? Either hit reply to this email or click here to leave your response*!"

Most people click the link and fill out the form, but some people reply to my email as well.

This information also helps inform me on what offers I can create for my audience and where they might be stuck.

Once you've got an idea of what's keeping your target audience awake at night, think through some of the amazing experiences that you've been through.

For example, remember when you first joined a membership site, or when you first joined a subscription? Has there been any business that stood out in terms of what they were doing and did to make you feel amazing?

Some awesome experiences I've had include:

ConvertKit

When I joined ConvertKit. After setting up my account and completing all the steps, they sent me an email congratulating me on this success.

Then they said that because I'd done everything they asked, I could get a t-shirt and care package at their cost. Who doesn't want that?!

ConvertKit continues to be amazing in the way that they serve. Their customer service is one of the best I've ever had access to. They are constantly helpful and always quick to respond to help requests.

I have been with ConvertKit for close to 4 years now and I have no desire to move anytime soon!

Niche Spa

This is a local brick & mortar business that my husband and I go to frequently to get hot stone massages and relaxing facials.

After our first visit, aside from the outstanding service provided during our visit, they followed up with an email to check in on how we were feeling and to offer us a 15% discount on our next booking.

They have also sent me a voucher on my birthday for $100 discount on any service, which is extremely generous.

They have this gorgeous waiting room that has big, comfy couches, luxurious pillows and lots of yummy treats to tuck into while you wait.

We love going there because it's so relaxing... exactly what we want!

Heartbeat

Another great experience upon joining. I was lucky enough to purchase lifetime access to Heartbeat. Meaning that I never have to pay for it ever again.

They use their own platform to help their users too. They have a space called *The Hearth*. They offer free workshops, support, and are receptive to feature requests..

Heartbeat is where my memberships, the *List Building Collective* and the *Digital Income Accelerator*, are located.

They sent me a ton of training videos and emails once I joined and continue to check in on my progress.

***

Hopefully this gives you some ideas as to why creating a WOW experience for your customers or clients is important.

Wowing your audience is how you will differentiate your business from somebody else in the same niche who's providing exactly the same product or service as you.

It's your time to shine!

WOW EXPERIENCE IDEAS

Struggling to come up with what might be a great way to WOW your audience?

Here are some ideas that you can get started with:

#1: Sending your brand new client or customer a thank you card in the physical mail.

I'm talking snail mail here. Now this means that you need to have their postal address. And if you're selling digital products or services, you aren't necessarily capturing that information.

But there is a way around that. There is a service called Postable.com that allows you to capture these details, plus more.

Obviously, it's optional for your audience to provide these details. But once someone has just purchased an offer from you, providing a way for them to get even more from you by providing a simple link to capture this data... well you'll find most people will fill this out.

This service allows you to send out cards via snail mail.

You'll get notified when it's their birthday, when it's their wedding anniversary, and anything else you ask about and they provide the answer to.

In some locations, they also include gift cards as well.

This is definitely an easy service to set up and WOW your new subscriber/customer/client right from day one.

#2: Send a coffee gift card.

You could also send out a coffee gift card after your new member has been in your membership for 14 days.

You can use something like Starbucks, or find out their local brew... there's loads of online options for doing this. You just gotta think outside the box.

#3: Send digital gift cards.

You could also send them a $5 Amazon gift card, or any other digital gift card. E-cards are amazing and easy to get into the hands of your ideal customer or client.

But we're looking to WOW our audience, so if you can send something physical in the mail, it will make you stand out. Your new customer or client will not be expecting this from you.

And when they do receive it, they will feel amazing and fall in love with you and your business.

So that takes care of when they become a new client or customer.

But what about after they have been in your world a while?

EXTENDING THE WOW EXPERIENCE

The goal is to set up a system that allows you to consistently WOW your clients and customers.

After the first 30 days, if you've done some of the options above, you will have cemented the relationship.

But what about creating something truly amazing?

What can you do?

Figure out the best milestones for continuing to WOW them.

How long does your service go for?

Your goal is to map out a timeline of WOW experiences during this time.

Let's take a course, for example. Let's say the course lasts for 12 months.

In the first 30 days, you'd send a card and Amazon e-gift card to a new member.

At the 90-day mark, you send them a coffee card and your fav sweet treat to congratulate them on their progress.

At the 6-month mark, you decide to shine a light on them via social media, telling everyone online how amazing they are.

At the 9-month mark, you gift them a 30-minute 1-on-1 coaching call with you to discuss their next steps and help get them unstuck.

And at the 12-month mark, you invite them to join your membership and give them a 30% discount for being a loyal customer.

You also send them a card with a little something special included.

To recap, it would look like this:

- First 30 days, send a Postable card and e-gift card

- 90-days, send coffee card

- 6-months, shout on social media

- 12-months, join membership with a 30%

discount

This is an awesome WOW experience and your customer or client is going to never want to leave.

You're going to surprise and delight them as often as you can.

It really is the little things that count.

So think about things that might make sense to do that are easy to implement and that don't cost you a ton of money.

And if you're selling a high end product or offer, like anything over $1,000 dollars for example, then it's worth it to send them some type of gift basket once they join, don't you agree?

Etsy is awesome for this type of thing. You can jump on *Etsy*, get something custom made for that client and shipped straight to them.

Remember, it's about creating an amazing, WOW experience from your ideal customer or clients perspective... not about what's the most beneficial to you.

Now that you've decided what your WOW experience is going to look like, let's figure out how to automate this as much as possible.

Step 2:

Set It On Autopilot

Before we jump into automating the WOW experience, it's important that you differentiate what you'll do for new leads versus what you'll do for your customers or clients.

In the previous step, we really focused on the customer/client experience, but you could apply something similar to your new subscribers, those that join your email list.

Particularly if you're looking to continue to attract your ideal clients to you.

We all know that your ideal leads have already received a digital gift of some description... So how can you extend the WOW experience to them without spending a ton of money on postage?

Here are some ideas:

1. Gift them a copy of your lowest paid product (ideally something that costs under $30) if they open your emails and engage with you within the first 30 days

2. Send them an e-gift card at the 90-day mark if they have opened 80% of your emails in the past 90 days

3. Gift them a complimentary 15-minute hot seat call with you at the 6-month mark

4. Shout them out on social media if you see them promoting your stuff (do this anytime)

Remember, it's the little things that count!

AUTOMATE YOUR WOW EXPERIENCE

Take the time to map out what your WOW experience looks like for each of your 'buckets' aka subscribers, clients, products etc.

Then decide on the milestones or frequency with which each action will take place.

For example, if you're working with a one-on-one coaching client, you might have a task to send them a card within 7 days of them signing up.

Then, you might have gifts assigned to go out at 90-days, 6 months and 12 months. Map that out in your calendar or fav to-do app.

Then, as you sign up a new coaching client or customer, you can go and add the tasks to your calendar or to-do app so that you get reminded of when to take action.

Do this for your subscribers as well. Except this process might be more of a review process for each milestone or frequency, where you check these monthly and take action accordingly.

Implement this phase and you'll really stand out to your ideal customers and clients and it's also how you'll attract the right leads and keep them on your list.

And it works. I've had a number of different customers or existing clients reach out to me who say, "Hey Lise, I knew it was right to work with you because you just sent me this amazing card..."

It's just that little thing that doesn't really cost you much but shows that you care.

Your task is to come up with some different ideas that might be a good fit for your business.

Choose what those milestones or frequency pieces are and then implement!

Create your milestone map and to-do items. Automate them using services like Dubsado.com or 17hats.com.

The point is to take action now. Set it up in your Google Calendar to get started if something like Dubsado isn't right for you now.

Don't use technology as an excuse to not do this step!

Your Key Takeaways:

1. You should have a list of WOW experience ideas ready to implement

2. A map of milestones for creating WOW opportunities for your new customers and leads

3. An automated process so that you don't forget to do any of these things!

And that's it. You've just finished *Part 3: Captivate.*

This was the final piece in how to *cultivate your audience.*

If you truly want an online business that excites you, that fills your soul, and serves the audience you care about, this process will help you achieve that.

It will also help you start to map out what consistent income might look like for your business.

But reading this book isn't enough. You need to take action.

In the next chapter, you'll learn how to take everything you've just learned and turn it into strategic action.

You got this. I believe in you. Your ideal customers and clients believe in you too.

Ready? Let's jump straight into your 30-Day Action Plan.

AMPLIFY, ENGAGE, CAPTIVATE

30-Day Action Plan

One of the biggest obstacles for most creative business owners in implementing these three components is that they don't have the right strategy or focus to begin with.

They don't know what to expect or where to start.

So instead of moving forward, they file it away for a later date.

That's why I want to provide you with a clear action plan on what to do next.

How to take what you've just learned and put it into practice.

We'll start by dividing a 30-day time frame into three 10-day phases.

Each phase will focus on either attracting, nurturing, or captivating your ideal audience.

It'll help you remove the roadblocks so you can stick to a plan and put it into place.

It will also allow you to identify where your gaps are and get started on filling those in.

But most of it all, it will allow you to create a plan for establishing real relationships with your ideal audience.

Building the know/like//trust factor so that you can transition into predictable income-generating months.

Days 1-10: Phase One - Attract

Over the next 10 days, you're going to focus on the pieces that will capture the attention of your ideal target audience... the people you really want to work with.

You'll figure out what your best lead magnet is, create it and set it up, ready to be delivered to your new subscribers.

Days 11-20: Phase Two - Engage

In this next 10-day phase, you'll focus on setting up your email service provider, creating sequences that convert your new leads into willing and long-term subscribers and ultimately, customers or clients.

Days 21-30: Phase Three - Captivate

In the last 10-days, you'll focus on creating the most amazing WOW experience for your leads, clients or customers. You'll spend time mapping out the milestones and setting up all the pieces to automate this process.

Get access to the 30-day action plan at here: https://www.hustleandgroove.com/cccplan.

A Final Note to the Reader

As we near the end of this book, I wanted to take a moment to tell you that you don't have to be perfect about this stuff.

The whole point of this book is to help you cultivate an audience that you can connect with on a regular basis.

To continue to develop real relationships.

To make an impact and change the lives of those you reach.

And this means letting go of things needing to be perfect or 'just right' before you implement them.

You might not think it's worthwhile creating a lead magnet. But if you want to attract the right audience, how do you plan on doing that?

Sending out regular emails might not be your thing, but how can you expect your ideal customer or client to know what you provide if you're not telling them about it?

Your job, as a creative business owner, is to create the things you're passionate about.

But if you want to be a profitable business owner, it means that you also have to step back and put your business hat on every so often.

By putting the right systems in place and automating as much of this process as possible, you can focus on creating while also working smarter and letting the tools available to you to do the heavy lifting.

At the end of the day, you know your business best. Do what *feels* easy and fun for you.

But don't do nothing.

I'm certain that if you take action, implement the 30-day action plan, and follow through on what you set out to do, you will become a lead attracting machine... but not attracting just any lead.

You'll be attracting the right lead.

Remember, it's far easier to make money from your existing subscribers, clients or customers, than it is to find new leads.

Focus on engaging, encouraging, and captivating the people you are already serving and your creative business will start to take on a very different feel.

I believe you can do this.

And if you're stuck, just take a deep breath and ask yourself this question: "What's my immediate next step?" And do that step.

The next step, as far as this book is concerned, is to grab the next book in the **Cultivate Your Business Series**, *Cultivate Your Cashflow: Create a Money-Making Marketing Plan Using Human Design.*

You can find that book and the rest of the books in the series here: www.hustleandgroove.com/cybseries

And if you're ready to attract and cultivate your audience every single day? You'll discover that and more inside the List Building Collective membership. You can learn all about it here: https://hustleandgroove.com/lbcbl

I can't wait to welcome you inside.

Take care

Lise xoxo

About the Author

Lise Cartwright is a bestselling author and creative business coach who is obsessed with helping others create and grow a business and life they love!

She loves curling up on a comfy couch with a good book, a hot cup of Chai Latte, and the soothing sounds of waves crashing against the white sandy beaches of the Gold Coast, Australia.

She's the founder of www.hustleandgroove.com, the #1 online resource for getting clear on your business model and growing an online business you are excited to work in. Her business motto is: "if it's not easy and fun, why do it?!"

Through her books, training videos, and coaching, she's helped thousands of people on their journey to creating an online business that's easy, fun, and profitable.

You can connect with Lise on the following social media platforms:

- facebook.com/hustleandgroove
- instagram.com/lisecartwrightnz
- linkedin.com/in/lisecartwright

Can You Help?

THANK
Y☺U!

If you enjoyed this book, could you PLEASE leave an honest review from where you purchased your copy from?

Reviews are really important to the success of a book — so if you like (or don't like!) what you've read, PLEASE take 2 minutes to leave your honest review — I really appreciate it.

www.ingramcontent.com/pod-product-compliance
Lightning Source LLC
Chambersburg PA
CBHW040928210326
41597CB00030B/5225